Shepherds' Crooks and Walking Sticks

Dalesman books of related interest:

COUNTRY CRAFTS IN YORKSHIRE
LIFE AND TRADITION IN THE LAKE DISTRICT
LIFE AND TRADITION IN THE YORKSHIRE DALES
SHEEPDOGS AT WORK
SHEPHERDS, SHEEP AND SHEEPDOGS
YORKSHIRE CRAFTSMEN AT WORK

Shepherds' Crooks and Walking Sticks

by

David Grant and Edward Hart

Dalesman Books
1985

THE DALESMAN PUBLISHING COMPANY LTD.
CLAPHAM, via Lancaster, LA2 8EB

First published 1972
Second edition 1975
Reprinted 1976, 1978, 1980, 1983
Third edition 1985

ISBN: 0 85206 842 5

Printed by Fretwell & Cox Ltd.
Goulbourne Street, Keighley, West Yorkshire BD21 1PZ

Contents

Foreword

by Lady Duncan

I HAVE ALWAYS admired the beauty and feel of a well-made shepherds' crook. At the Royal Highland Show in 1965 I stood watching David Grant at work with his sticks at the Scottish Crook Makers' Association's stand and asked him where I could get a book on this fascinating craft; he replied that there was no such book but he hoped that some day there would be. Here is the book, carefully compiled by Edward Hart from the great knowledge of David Grant.

Having attended the winter evening classes at Lockerbie on stick making, at which David Grant was the instructor, I know how much pleasure and satisfaction there is in this craft. Many of his pupils will have made sticks of which they are proud, thanks to the skill and patience of David Grant. It is easy to see by looking at the exhibits of sticks at the various agricultural shows that they are the work of highly skilled craftsmen — Dumfriesshire is fortunate to have so many.

This craft can be passed on to each generation best by the teaching offered at evening classes by such men as David Grant. This splendid book will ensure that the art of stick and crook making will never be forgotten.

Isabel M. Duncan

Castlehill,
Kirkmahoe,
Dumfriesshire.

Introduction

THIS BOOK concerns the art and practice of stick making. It includes walking sticks plain and fancy, shepherds' crooks, and ornaments in wood and horn. Although we realise that no book can replace tuition under an expert, many enthusiasts live either in the hills far from classes, or in urban centres which do not cater for this particular interest. Basic principles outlined here should help them, as we believe it will those fortunate to attend the admirable evening classes run by county councils. We are indebted to Yorkshire farmer Mr. Frank Graham for his considerable help on carving ornaments, a branch in which he has few peers. Mr. Graham's forebears also came from the Borders.

The practical aspect of the stick and crook is always borne in mind. Each must be so balanced as to be of positive help to the man walking the hills. It is a matter of surprise to us that town visitors walk for miles without the aid of the "third leg," without which no professional shepherd would venture 20 yards! A good stick can add miles to a day's accomplishment, and ease any journey on foot.

To make the properly balanced stick or crook, whether for use or ornament, has proved an abiding source of satisfaction to men and women in all walks of life. If this book helps the continuance and spread of the craft, it will have been well worth while.

1. History

THE CRAFT of stick making is very, very old. Carved sticks have been found in ancient tombs opened up in various countries, though the shepherd's crook seems to be peculiarly British. How did it arise? Certainly it goes back for generations, as the oldest man to whom I talked as a boy had no recollections of the time before stick making began, or had heard of such a period.

Half a century ago, shepherds were almost the only crook makers. We may reasonably assume that they had always carried a stick of some sort, to aid their long, uphill treks. A small branch at the end could have been shortened to catch a sheep round the neck, after which it was a natural step to take a curly ram's horn found on the hill and, after the roughest shaping, attach it to the end of the stick.

Shepherds' cottages of those days were primitive affairs, often with nothing better than a paraffin lamp for lighting. This was barely sufficient for reading, yet adequate to see to carve a crook. It had the further advantage that heat from it could be used to bend the horn. In those days before wireless, television, or a daily paper in outlying parts, mid-winter evenings must have seemed long indeed. The hill shepherd whiled away the time by whittling at his crook.

Today, things are very different. Shepherds may still make their own sticks, but they have been joined by a great number for whom stick dressing is a fascinating hobby, a shaping of natural materials for use and ornament, combined. It is, if you like, a reaction against mass production. One batch of plastic is exactly like the last in colour and texture; no two hazel or holly shanks are identical, and certainly no two horns are alike after they have roamed the high hills of England, Scotland and Wales on the heads of flock leaders of wandering and sometimes quarrelsome disposition.

Each horn and each shank present a different problem. This is why evening classes on the subject are filled. This is why their spread would be so popular, encouraging a hobby that helps many and hurts none. Last winter at Lockerbie I had 32 budding stick makers, so many that the class was split in two. Dumfriesshire County Council deserves thanks for its sponsorship, yet must feel rewarded when the end-of-sessions display is exhibited.

The greatest men of our day have been among those thousands fascinated by the endless variety of sticks and crooks. Sir Winston Churchill's collection may be seen at Chartwell. Every country house has its selection, and the Archbishop of Canterbury's ceremonial staff is a crook.

Shortage of good horn is indeed a limiting factor, but wooden sticks are also most pleasing. There are still hazel copses, hedgerow holly, and thorn in abundance, though if conifer planting continues on its present scale even these hereditary assets will be denied us.

2. Cutting a Shank

"When is the best time to cut a stick?"
"When thoo sees t'booger!"

THE YORKSHIRE dalesman tendering such down-to-earth advice had obviously experienced the frustrations of choosing a good stick, marking it and watching it grow, and then finding it harvested by someone else! Though this has not happened to me, I have more than once noted a particularly fine specimen growing in a wood, and then at the appropriate time been unable to find it.

Thus our dalesman friend's advice has much to commend it. If you can be sure of being able to cut at the correct season, wait till the leaves fall. This may be in October, but certainly from November until the end of January is the ideal period. Our object is to start with the minimum sap, so cutting takes place in the "dead" season. Naturally, dates vary according to district; sap will rise in Devon's deep valleys some days before it starts on these Border hills. If the stick must be cut while the sap is rising, set the block in running water for a month or six weeks to help removal of sap. I have done this successfully on several occasions, for any season will do for me if the stick is really good.

The crooks and sticks we are dealing with may be divided into four categories. These are horn crook, wood crook, walking stick (quite different from a crook) and thumb stick so popular in the north of England. Among this list is a wide variety of design and materials. Horn will be dealt with later, but the best horn requires a good wooden shank just as does the wood crook or stick.

The British Isles provide abundant choice. Hazel is most common, being found in many kinds. White hazel is inclined to be thicker in the bark than dark hazel, so the latter requires a finer sandpaper when you reach that stage. Other useful sticks are cut from ash and holly, and from both of these the bark should be peeled when cut. Blackthorn is another good one, though care should be taken when removing the knots. Sharp lumps must not be left, but the stick looks more attractive when a certain amount of knot is retained. Wild cherry is used, and some people like to carve unusual sticks such as plum. A variety of willow locally termed hack berry, grows in wet places, and I have used it. For making a wooden crook or handle to fit on another shank, the hard knot or burr growing on the trunk of an ancient elm is unbeatable. Beech is also good for this purpose.

Even for a horn stick, the choice of shank is vital. Some stick makers are "horn minded"; they will take a saw and whip the shank off a good wood head, with no consideration for the use that a wooden crook maker might have got from it. A lot of good sticks are wasted in this way. Others care little about the shanks so long as the horn may be carved artistically. This is wrong. The shank is a very important part of any stick, and not something to be considered as secondary.

I look for a natural neck in a wood stick. A ewe charging down a hillside at full gallop exerts considerable force, and all the weight is brought to bear on the neck of the crook when she is pulled up. This is always foremost in my thoughts; the most fancy crook ever carved is of no use to me if it could not be used to catch a Blackface or Cheviot in full flight. To find this neck, a spade is needed. Dig down and take as much of the root as necessary. A golden rule in stick making is that you can always take more off, but you cannot put any back. This applies equally to the initial harvest and the final smoothing.

Other tools for cutting your shanks are a good Bushman saw and a cross-cut hand saw. Select a shank of the required thickness and as straight as possible. Any with "dog legs" must be avoided, but other types of bend can be taken out with heat.

Another type of one-piece wood stick has a shank growing from a thicker part of the tree or bush. The main trunk carrying your favoured shank should be sawn off four or five feet above the junction. This is to prevent any splintering from running down and spoiling what is to become the head of a wooden stick. I have ruined good sticks in this way before I learnt the trick.

Measure your shank length hand over hand, till you have a stick 14 hands long. It is a good guide, although the human hand varies between individuals. Length is really a matter of taste, but crooks should not be so long that the handler lose contact with the sheep. He must catch it with his spare hand. The further north you go, the longer the stick as a rule. Highland lairds like to have something on which to lean at a mart!

Each trunk is now sawn to the length shown in photograph 5 to leave an ample block. Take these home and stand them on an uninsulated cement floor. Damp will strike upwards and prevent too rapid drying out. Blocks should be levelled and dressed as much as possible at this stage. Shanks for horn sticks have no block. They dry best when tied in bundles of about 12. Cut a good, straight, thicker one for the centre, and build up the bundle of shanks round it, tying both ends firmly. Then hang up to dry.

Individual sticks may be hung, head uppermost, with a weight on the other end, but if you need a lot this method is impractical. Once in store, they must still be watched. If any cracks appear in the block, pour in linseed oil. Two years is the minimum drying period, and with a big block it may be three years.

Regarding wooden blocks, you will naturally look closely at every piece of timber which you plan to saw off for a stick head. Unfortunately, the soundest looking block may reveal unexpected flaws in the working, at the very point where you sought perfection. This is another reason for having plenty of choice on hand.

Be sure to label each bundle of sticks with date of cutting, otherwise you will be in a mix as time passes.

3. Selecting the Horn

SELECTING HORN is a rather optimistic phrase. The stick dresser must work with what he can get, often using horn with various sorts of flaws, simply because nothing better is available.

Abbatoirs are the likeliest source. This entails a special journey, and considerable payment for a single horn, an expense often forgotten by stick purchasers. They tend to think that the carver has had no cost save his own time in his own workshop. Hill farmer friends can help, but of course both they and their shepherds may be stick dressers, and pounce immediately on any horn should a tup die.

Farmers can help in another way, and that is by taking care when they have to bend horn which is growing into the sheep's cheek. Formerly, a swede turnip was boiled and thrust onto the horn, applying the necessary heat evenly. No-one has time to boil a turnip today, so out comes the blow lamp, giving a quite unsuitable type of heat.

A fairly large horn is needed to make a crook, whereas for a walking stick a smaller horn will do. Usual types are sheep, goat, buffalo, cow and stag. Since stick dressing became more popular, we have learned that horn from almost any animal can be used, but that all have different characteristics. For instance, stag horn cannot be heated and fashioned; I use it chiefly for ferrules, as explained in chapter 8.

Ram's horn is our main material. A good clean horn is sought, and by that I mean one that is free from cracks, and fine-grained. Bumps and bulges can be taken off, and hollows filled in to a certain extent, but size is important. "Plenty of cloth" is the stick maker's aim in horn selection, as he then has something to work on. Thickness of actual horn at the base is important.

Regarding breeds, Scottish Blackface is both good and fairly plentiful in many hill areas. Swaledale is very nice if there is plenty of horn to work on, though inclined to be a bit thin. I have made sticks from Dorset Horn, which took a lot more working owing to the double curl. Some Cheviot tups are horned, and are quite suitable. I have never used Herdwick nor Welsh, though they should be quite all right unless too small.

A Jacob sheep breeder sent up a pair of horns from the Royal. I have tried them recently and find them satisfactory, never having had the opportunity before. Jacobs are very popular at the moment, and may help offset the large numbers of hill tups lost through afforestation. Lonk is also a useful horn.

With any sheep, the older the horn the better it is for crook making. Every extra year tends to lengthen the horn, and shearlings are of use only if the horn is long and thick enough. Aged tups do tend to have horn damaged through fighting or flies, but generally there is nothing to beat them for quality and texture.

I came across my first buffalo horn in rather a queer way. My son, Sandy, fished it out of the river Annan, into which it had probably been thrown when someone was spring cleaning. The horn was over a yard long, and made a very nice stick.

Remember that the whole basic difference between a show stick and a working model is in the horn. Horns damaged by flies or fighting may be perfectly serviceable to fashion into a head used for walking the hills, and holding an escaping ewe as she comes over a rocky ledge, but will never be flawless enough to catch the judge's eye.

The best looking horn can hold hidden snags. This is a disappointment for which every stick maker must be prepared. Some fault or weakness occurs inside which could not possibly be spotted from the outside. It adds to the joys of stick making when a sound horn works in the way you had hoped, and comes up into a lovely finish without any unsuspected flaws. Horns are almost everlasting, so there is no need to store them in rotation.

Be sure to label each bundle of sticks with date of cutting, otherwise you will be in a mix as time passes.

Rams' horns. Those in the top picture are long enough but are light and shelly. The horn in the bottom picture is good and solid, and provides plenty of scope for working.

4. Tools

ONE POINT should always be borne in mind concerning tools. Many a good stick has been fashioned simply by whittling away with a pen knife. Other equipment is used to speed the task, rather than to improve the finished work. Like everyone else, stick makers must move with the times, and some of the latest tools are a great asset.

For cutting sticks in wood or hedgerow, we have already mentioned spade and saw. A light spade with a narrow spit is best, like those used by rabbiting men; it is handy to carry. A Bushman saw has many advantages, but is awkward to use at ground level. In that situation a good handsaw is better. I prefer the medium axe with 4lb. head, being fairly light to carry, yet capable of severing roots easily, if directed properly!

Two items essential to the keen stick maker are a bench and a vice. The former is likely to be what one can find and has room for, but a vice should be chosen with care. Visit a farm sale, and try to buy an old one. A blacksmith's vice, or something similar, is best; it will be of far better stuff than a new one, which contains a lot of cast metal. A vice that has withstood years of work has a recommendation behind it, and will serve for many more seasons to aid the stick dresser. Wooden vices are useful, especially if they are set flush with the side of the bench. Shanks can then be straightened in them.

Regarding specialist tools, a useful foundation may be built up for about £25, adding refinements as desired. Saws, files, drill, glue and smoothing materials are the basics. A lamp with a funnel is also needed. Aluminium is best, and glass should be avoided, as it is always liable to be knocked off and broken.

A wide variety of files is a source of satisfaction, as the exact type for the job may be picked out. Files are available in different sizes and different grades from coarse to fine. Very small, round files are used for getting into all the corners when doing fancy work. Surform files may be bought with flat, bevelled or cylindrical surfaces, replaceable on a standard frame. They are a great help to the stick maker, as are files with the different variations on the one tool. A favourite of mine, shaped like a sharpening steel, was purchased in Woolworth's 15 years ago. Finer files are liable to break, so three or four of each type are advisable.

For actually dressing the stick, three types of saw are available in the frame class. A fret saw is for working round about, in very fine corners. A

A selection of tools used in the making of shepherds' crooks and walking sticks.

bow saw is useful for cutting out the wooden head, while a coping saw is intermediate. A rip saw cuts with the grain, and a cross-cut at right angles to it. You will find a sharp rip saw best for cutting the sides off a dried root or block.

5. Wooden Block Sticks

WE HAVE now reached the stage where our wood blocks with shanks attached have been drying for two or possibly three years, and arrive at the actual making of the wooden stick.

Job number one is to straighten the shank. To do this, it is heated at any bends with dry heat from a lamp. A sound precaution here is to wrap the parts to be heated in silver foil smuggled from the kitchen. If you put aside a sheet now and then, the cook will never miss it! This device prevents burning the bark. Straighten the shank on your knee, or put it in a vice against the straight edge of the bench. Your shank must be absolutely dead in line, or you will never have a first class stick.

Next move is to take the sides off the head. Use the most appropriate type of saw, as described in chapter four on Tools. Above all, do not cut the heads too narrow, as it is a simple matter to shave away a litle more if necessary. You are left with a straight shank and a flat, unshaped head. Draw in pencil on one side the shape of the head, allowing three and a half to four inches between nose and inside of heel for a crook — less for a walking stick. For a block crook or stick, a "natural neck" is preferred, one in which the natural joint of the shank to the block continues in unbroken curve. For strength in the finished article, a natural neck is best every time. I can just get my hand through the gap in the case of a crook, but human hands vary, so this guide must be checked in each individual.

When drawing the outline, always mark the inside first. The outside line then follows of its own accord, but if you start with the outside you are sure to get the inner cut wrong. This applies to any type of stick or crook.

Cut round the inside mark, using a bow saw. Now use Surform file or pocket knife to roughly round off the edges. At all times you must bear in mind that there is nothing flat about a stick head; it may have an oval section, but never a square one. It must be made to fit your hand. Carve off this inside before taking off the top, and at all times try to preserve as much bark as possible, leaving it evenly on both sides. Make the crook higher at the nose than at the heel. This is absolutely vital, and applies equally to a walking stick. A head with a drooping nose will never look right.

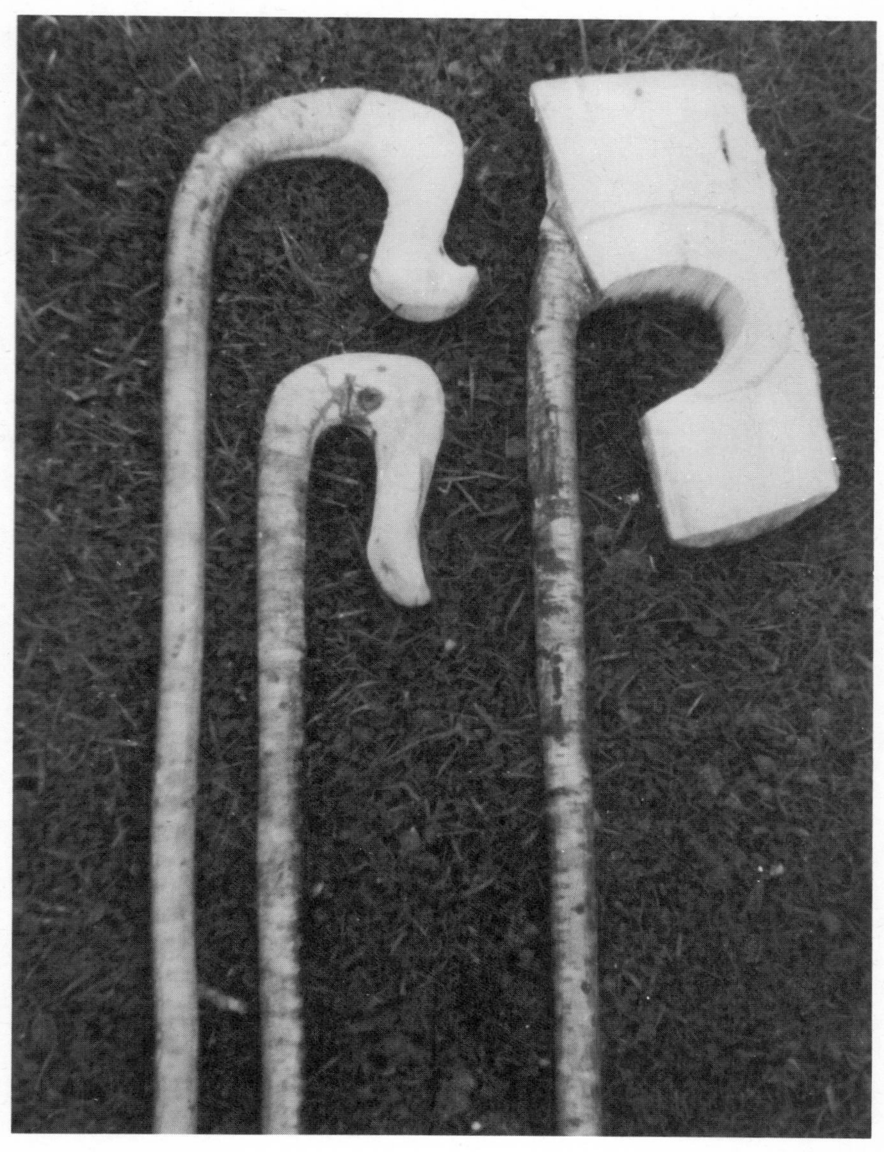

All-wood sticks and crooks. Early stage (right) and further shaping (left).

All-wood sticks and crooks. Hazel for the one-piece stick cut from the main stem (left) and including some root (centre).

6. Shaping Horn — First Steps

WE RETURN to our horn to discover whether it may be fashioned into a thing of lasting beauty, or whether it has hidden flaws. These are found only during the the working of the horn, for which very definite rules have been evolved.

First stage is to draw a pencil line across or round the base of the horn. It may be already nicely squared, but if not, then this must be done. At this stage, saw off the minimum to give a right angle with the proposed heel. I cannot over-emphasise that you must start with the heel. Set the horn in a vice, and file the back of the heel as straight up and down as possible, using a Surform or file. You may also have to remove some rough horn from the inside.

At this stage, the horn still has its typical ram's curl. It must be turned, starting with the heel. This is vital, and may only be done while hot. Before heating, reset your cold horn in the vice, and decide which way you will turn it. There is no point in having it thoroughly warmed and then wondering what to do with it! Then leave your vice set ready, so that you are not desperately screwing up while losing heat.

Bends are taken out from the heel, working towards the nose. Heat your horn by holding over the funnel of an oil lamp, turning it as needs be so that every side of the part to be turned is evenly heated. You will soon learn the amount of heat needed; as a very rough guide, it is not so hot that you cannot touch it, but is too warm to hold in the hand for many seconds. Tin foil may be used to prevent burning and retain heat. You should, however, grasp it momentarily in your hand to check this necessary evenness of heating. There is no standard temperature that I know of in degrees fahrenheit to help, or even in degress centigrade.

Heating by boiling is possible, but when heated by lamp the horn has much better chance of keeping its shape. Remember, horn and wood will always try to revert to the shapes in which they grew. Boiling a big, thick horn gives more working material when flattened in clamps. As stated elsewhere, on no account should you use a blow lamp, for the severe heat spells death to any horn.

When you think that the correct degree of heating has been applied, set the horn in the vice with the heel perpendicular. Wind up the vice, ensuring that the horn is not twisting from one side to another. Look at it from all

Rough shaping of horn for walking stick heads and leg crooks (lower row).

angles, keeping the head fairly level. Take out the bend by using a pair of grips on the nose, or a metal cylinder. Lever back to the required angle, then tie grips or cylinder firmly in the desired place. You should have ensured beforehand that some simple fastening point is available, and if not, make one with staples or hook.

Wire or nylon cord is good for tying back. Nylon does not give, and is cheap and readily available. Though stick making is an ancient craft, that is no reason for failing to use modern devices. There is still ample scope to task the carver's ingenuity without sticking to primitive methods. Two stick makers at Canonbie Show found that wire was the only thing that would hold the very strong buffalo horns on which they were working.

Having tied the reshaped horn securely in place, you must wait for it to cool. Leave it overnight. Do not attempt to cool it with cold water, or it will tend to return to former shape. Some of my pupils may say that I do not practise what I preach here. We did cool heated horn with a wet rag at our Lockerbie classes, but this was simply to speed up the demonstration process, as I pointed out to my pupils at the time, if they were listening! If you must cool in this way, a pail of clean cold water into which the cloth is dipped from time to time is all that is needed.

When the horn is thoroughly cool, decide on further straightening. The nose must be plumb in line with the heel when the head is turned. Turn the horn as thick as possible, as you then have a wider scope on which to work. Two or three heatings and straightenings may be needed.

I am looking for a paraffin stove with an elongated rather than a round funnel, to apply heat over a greater area of horn. All sorts of gadgets may be developed to aid stick making, but we don't want to frighten off beginners by making the art appear unnecessarily complicated. Always keep in your mind those early shepherds with their lamps, penknife, and precious little else except time and skill. Patience and perseverance are essential. No good stick was ever made in a hurry. To say "I made this stick in two hours less time than the last" is no standard at all.

When the nose is perpendicular with the heel, the next stage is to shape the inside. As in the case of wood sticks, it is absolutely essential to shape the inside first, and then the outer line follows naturally. Don't waste time experimenting with this; I have tried shaping the outside first in my stick making's early days, and found it to be quite useless.

Mark in pencil the curve which you think will give a comfortable grip to the hand for a walking stick, and also be suitable for a sheep's neck in the case of a crook. Saw out the inside, using a coping saw. A pungent smell wafts up your nostrils when sawing horn. I can't say I care for it, but what does that matter if I get a good stick? After sawing, smooth the inside by using a cylindrical Surform. Now and at all times, try to keep clear of the "white," which is the softer, less mature part of the horn.

You can now see how things are turning out. To get the exact width of the head, further heat may be needed. Do not apply to the back, only to the sides and inside. After heating, lay the horn flat in the vice, having set that indispensable tool beforehand. I place a section of ash or oak inside the head, and tighten up onto it. My block for this purpose is 2⅝ inches long, and 1½ inches across to the top of the curve.

Flint or pitch is the inside support, containing blood vessels, around which the horn grows. It is best seen in the short stubs on the head of a new-born lamb of a horned breed, which are quite soft. Slight cracks may appear on the back during heating but but are usually only skin deep, and file off.

7. Shanking

AS THE amount of working material becomes more obvious, sketch the outline of the finished horn. If you want to hide the white, inner portion, simply take off the sharp edge which is a common feature in rams' horns. In the making of heads, whether of horn or wood, the inside must **always** be shaped first. The outside line then becomes apparent, but it is quite impossible to finish with a nicely curved inside line if you do the outside first. The heel of a crook may be either square or round. The round heel has a continuous curve, while there is a definite change of direction in the case of the square heel.

Shanking the head is the most vital stage in stick making. If this is not done correctly, the whole work is spoilt. Shanking, as we term the fitting of head to shank, controls the angle of the nose. A head should never droop. To shank a head, it must be bored to fit the shank peg exactly. Some stick makers bore before shaping the head, but I prefer to shape first. How far you take this shaping depends on you own preference, but in any event all horns must be fitted to the shank before the finishing stage. If this is not done, you are sure to find some irretrievable mistake through taking off too much.

A half-inch bit is the most usual size for boring, but may be larger or smaller according to the amount of horn available on the head. A mark should be drawn up the side, and possibly up the back of the heel to guide the bit. I then place the head exactly upside down in the vice, and bore downwards. Care in this operation cannot be over-emphasised, as it controls the whole set of head on shank. I know one man who presses the brace into his diaphragm, twisting with one hand and holding the horn in the other. He is adept enough at it, but this method is only for the highly skilled. Depth of boring depends on type of horn. Go as far as you dare without weakening the case of the horn in any way. Stop about one inch below the heel.

Now cut the shank peg. Measure from the head of the shank a distance equal to the length of the hole bored and cut round with a penknife. The peg must be shaped to the exact diameter of the hole, fitting easily. Keep trying it to make sure you don't remove too much.

If you wish you may strengthen the peg by inserting a steel rod inside it. If you do this, leave shaping the peg until the rod is in position. Another alternative is to use a metal peg without a supporting wooden one, in which case the steel should be about the width of an eight-inch nail. A nail will not do, however. Sound steel is essential.

Shank lengths vary, as we have discussed in Chapter 2. You may cut the shank to its proper length before making the peg if you wish, but I prefer the peg first. You then have something to work on if you make a mess of things. Always choose the shank according to the thickness of the head. After a horn is shaped up, you can gain a better idea of the type of shank required. It must be strong enough to carry the head adequately, but should be neither too heavy nor clumsy.

This is the reason why a large stock of shanks is needed. With a wide choice, you may be sure to find one that is just right for the particular head you are fashioning. At least three times as many shanks as heads are needed on hand, of varying length and thickness. To cut six shanks because you plan to make six crooks is completely futile, and remember that good shanks are usually far easier to come by than are sound horns.

A smaller peg will suffice for a walking stick than for a working crook. For those not experienced with sheep, I must emphasise once more the tremendous pressure exerted by a hill ewe going down a hillside in full flight, and all this is taken by the head of the crook and then by the peg. Your shanks should not have too much taper, or they will never handle correctly. A nice cylindrical shank balances itself in the hand with the head in place. A man walking 20 miles of rough country needs a stick as well balanced as a first-class gun or cricket bat, not something that will drag.

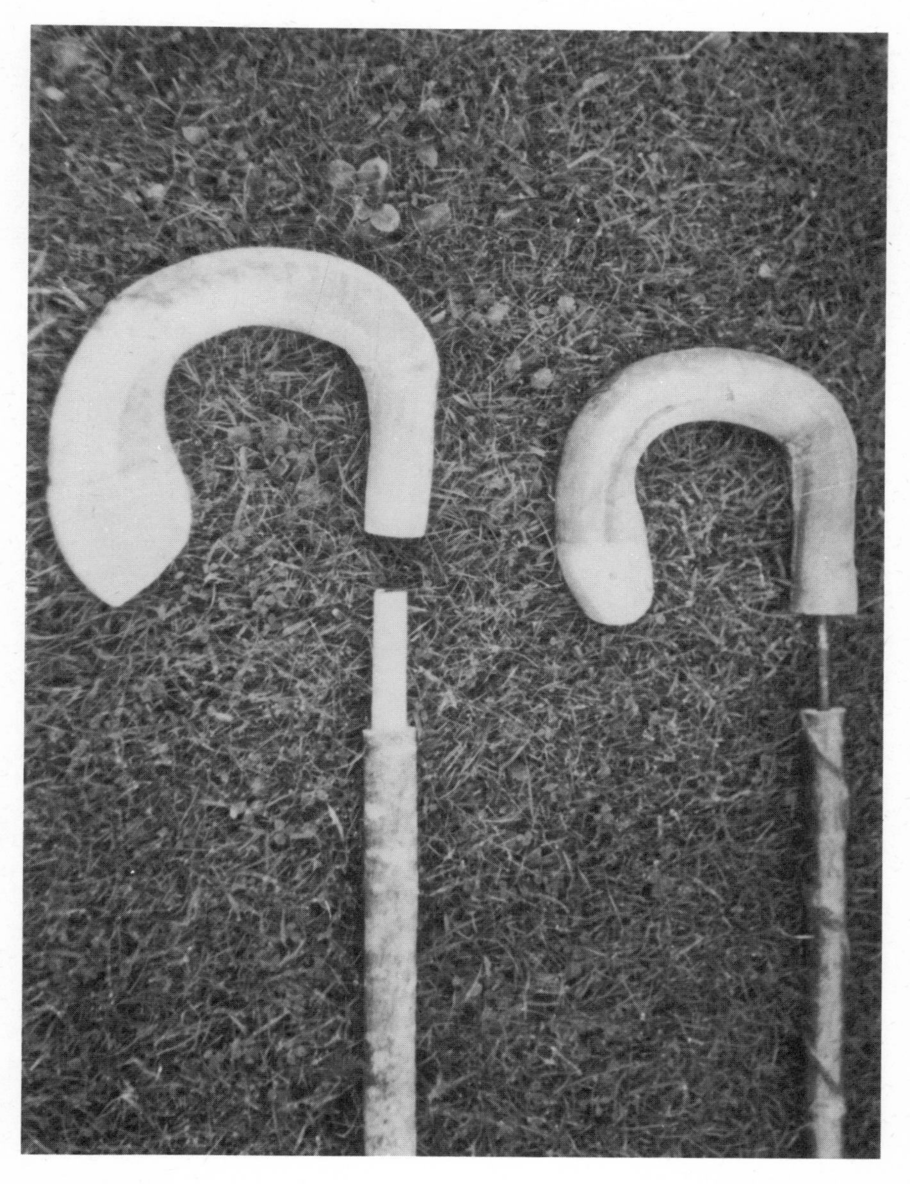

Shanking. The steel pin (right) and the shank itself are shaped to fit the head, which is then flattened at the nose so as to carve a scroll or thistle (left).

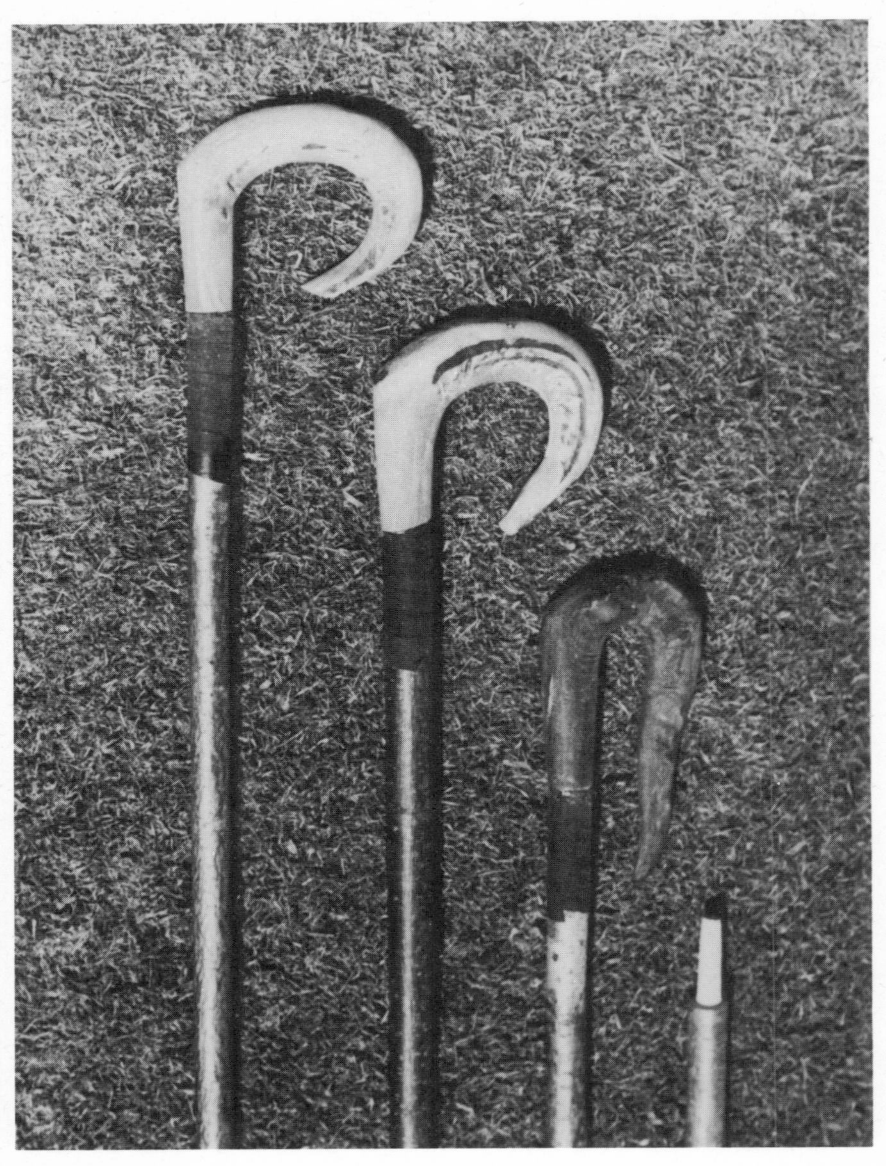

Protective binding after shanking. Note that the heads are not finished at this stage.

8. Tips and Ferrules

FERRULES IN stick making are of two types, and so cause some confusion. The one joins head and shank, the other tips the base of the shank to give longer wear. Both types are important parts of the properly finished stick. The tip is fitted last, but not so the ferrule. The purpose of the ferrule is to strengthen the joint between the head and the shank, whether the former be of horn or wood from a separate block. Some carvers aver that the stick is stronger if the ferrule is placed on the outside, i.e. not let in or countersunk. I maintain that a show stick looks neater with its ferrule set in, thus maintaining a straight line from tip of shank to neck of the head.

Ferrule may also be termed "ferrel," from the Scottish "virl." This is derived from the Old French word "virole," from the Latin "viriola," meaning a bracelet. Here we have a very good indication of the ferrule's place in the crook maker's art, for a ferrule can be a thing of beauty, the final touch of craftsmanship. I prefer to make the ferrule before the head is finished.

Various materials may be used. Stag and cow horn is as serviceable as sheep horn, while for a working stick brass, copper and aluminium are all favoured.

To make a ferrule, first cut your material to the length required, and as square as possible. Bore out the middle using the appropriate bit according to whether you are fitting the ferrule outside, or letting it in. Obviously, a smaller bore is needed if the ferrule is to be countersunk.

When working at it, fit the ferrule onto a bit of old shank set in a vice. File it down to the requisite barrel shape, using first fairly rough and graduating to smoother files. The barrel shape is both neater and tougher than a true cylinder, being strongest in the middle where most pressure is exerted.

When finishing, all file marks must be taken out. Final touches are identical to those described in Chapter 9. Before glueing, you must be absolutely sure that the fitting on both shank and head is accurate. Once it took me half a day to fit a particular ferrule on a show stick; on other occasions it has been dead right first time.

Brass ferrules may be bought for tipping. Quite a number of rubber ferrules are now used, especially by lame people to help prevent slipping. You can make a ferrule of horn, and fit it by boring the horn with a slightly smaller bit than the diameter of the shank, then fit it in exactly the same way as fitting head to shank. Leave enough horn to file some off, to get a nice round taper. This is very artistic, besides filling the tip's main function, which is to prevent splintering at the end of the shank.

9. Finishing

WE NOW come to the final processes, when all the natural beauty of head and shank is brought out. The stick is properly shanked up and the rough, outside horn taken away. Most of our hand tools have been brought into play to reach this stage. We now turn to finest sandpaper, glass paper and a soft cloth. Plenty of elbow grease is essential at this vital stage. Work away with your sandpaper until the horn is absolutely smooth. The rich natural gloss is produced after perhaps hours of work, and to enhance it, some stick makers use Vim or Brasso. These are rubbed on with a fine cloth, followed by another polish with soft, dry cloth.

I have shown many a stick that had no varnish on the head, but one has to keep up with the times. Varnishing is so popular with so many makers — and judges — that it may be necessary if a prize is sought. Personally, I prefer the well-finished stick with no vanish on it at all.

A tremendous number of different types of polyurethane are available, but some are definitely better than others for our purpose. Shellac varnish is good. It is bought in flakes, put in a jar, and mixed with methylated spirits. Shell of orange is a favourite colour, but the varnish may be dark, light or natural. Cottonwool saturated in the varnish, and then wrapped in fine muslin, is best for applying. Avoid using anything hairy, or you will spoil the effect completely. One or two drops of linseed oil rubbed over the pad prevent it from sticking.

The principle of shellac varnishing is the same as for French polishing, only instead of working round and round, the application must be given one way only. Several coats are needed, with thorough drying between each. Be very careful to have an even finish; it is all too easy to get a deeper colour in one place than in another. This applies especially to a wood stick, as the varnish is liable to sink in more according to the grain of the wood in different places. In some cases, a sealer is needed. You must keep applying it, and rubbing it off with sandpaper, until a base for the varnish is obtained.

A wooden stick is treated exactly the same on both head and shank. Some like a darker shank, and this is acceptable. Mahogany, dark oak and light oak are among the variations. Old-fashioned makers used flour, but good though this was, it has been superseded by other materials which make the stick dresser's difficult task a little easier.

You may not use paint, except in a fancy class. In these, it is a case of no

Finished "natural neck" crooks.

holds barred and the dresser is allowed any materials he chooses. Remember that the stick's prime purpose is as an aid to walking and catching sheep, and that no amount of fancy work can compensate for inability in that direction. I well remember an old Scottish shepherd after he had had a dram or two: "None of your light, fancy, show sticks for me," he said. "I want something that will help me up on the brae!"

Another shepherd who gave me some good advice was Mr. Adam Scott, who judged the first two sticks I ever entered. I was not among the tickets. Afterwards, this kindly man took me aside to tell me that the sticks had too much wood; they were too thick in the handle, otherwise they were good sticks. I fined them down a bit, and entered them at Langholm under a different judge altogether. They were placed first and second; they were the start of my career.

Some people like to carve a name on their sticks. I was never fond of this practice. It appears a little ostentatious, and is no longer as popular as it was in the early 1960s. Also, the practice of name carving has its dangers. A well-

Sticks of many kinds. Two finished horn sticks, a fancy stick with a fish head — and a ram's horn.

known Borders stick maker lost his crook, and it turned up in a lady's bedroom! Let that be a severe warning to all stick dressers. If the crook hadn't had his name on, no one would have known anything about the incident! If you must carve your name, however, a sharp penknife is the best tool. Experts maintain that horn should be left for the letters; i.e. the space around the letters should be taken out, and not vice versa with the letter themselves sunk in. The background is then painted red or black; I have seen letters set in red that looked excellent.

When finishing a shank, all knots are taken off. This must not be done haphazardly, however; the knot should follow the shape of the shank, and be curved to the same degree as the round of the stick. A great deal of charm is lost if they are dressed off level. A fine file is the tool for final knot shaping. For the shank itself, careful judgment is needed. A thick barn needs rough sandpaper to get the best effects, whereas the same grade would ruin a fine-barked shank. The finer the bark the finer the sandpaper used on it.

Hazels grow in different varieties. Some are much lighter coloured than others, and these lighter ones often have a spotted effect, which should be preserved and cherished as a covering of charm. If you are too heavy-handed with the wrong grade of sandpaper, this mottled formation will be permanently spoiled.

10. Fancy Sticks

STICKS LEND themselves to unlimited decoration. The main essential is plenty of wood or horn on which to work, and then you may put on any design you wish. Thistle and scroll on the nose are particularly popular north of the Border, while a fish handle, with fish's head protruding over the heel, is another favourite. Eyes may be bought for this purpose.

Taking the scroll or thistle as an example, you must first of all flatten the nose in a vice, to get as much horn with which to work as possible. Draw the required design in pencil, and whittle away with a penknife after cutting the outline with a fretsaw. Some carvers prefer the thistle or scroll to clear the nose; others join it up by leaving a small amount of horn uncut. I have on occasion made a crook in the reverse way, with the shank fitted to the nose of the horn instead of to the heel. I have a Cheviot horn worked that way at the moment, my idea being to gain more horn and so carve a better thistle.

Other popular forms include a sheep-dog coming up the heel, and sometimes another one creeping the opposite way, up the nose. Between them is a sheep being penned on the middle of the head. Pheasants, snipe, curlew, flowers and rose buds have all been carved onto shepherds' crooks.

Very many hours have been spent in carving these intricate designs. I greatly admire them, though personally the really well-balanced plain stick or crook holds greater appeal. This is not to decry fancy work in any way, but to my mind it must never interfere with the stick's main purpose.

Shanks are painted in the fancy stick classes, otherwise varnishing is all that is allowed. A wood stick may be carved in the same way as horn, fashioning birds, fish, animals or flowers from the block. I might add that in all cases of fancy embroidery, the appendages are **always** carved from the block or horn in one piece with the rest of the head, not stuck on afterwards.

Carving ornaments can never be learnt from a book. Each designer must decide on his own model, whittling away to copy the object, photograph or picture in his mind's eye to the best of his ability. The cardinal rule is not to cut too deep or too rapidly before checking; wood or horn can be taken off, but cannot be replaced.

11. Showing and Judging

YOU MAY take up stick making for your own amusement, with no intention whatever of competition work. Once bitten by the bug, however, the sight of well-filled classes at local shows may arouse the feeling, "My last one with its mottled hazel shank was as good as some of these." To carve a stick specially for showing is then but a short step.

Show schedules are obtained from local secretaries, and often include novice classes. You must study the stipulations in specialist classes, such as plain horn, or ornamental. Some painting may be allowed in the latter; in the others, never. Remember, there are judges AND judges! Some have never made a stick in their lives, while others are right at the top as craftsmen themselves.

For showing, every stick **must** be well finished. Final details count for much, but the nose must always be plumb in the centre, otherwise the stick has no chance at all, however ornate or polished. Judges vary in likes and dislikes. The great character, Mr. Ben Wilson of Troloss, could not abide a snake at any price. It was just a waste of time to put a serpent in front of Ben. Sadly, he is no longer with us, and sheep breeders, sheep-dog men and crook makers are all the poorer.

I prefer a good plain stick. This is not purely personal fancy; I have made it my business to stand and listen to what the general public has to say at shows, and there is no doubt about their choice.

Care of sticks at shows leaves much to be desired. They should be in stands with plenty of room, but sometimes they are clattered onto a bench in a heap. After the scores of hours that have gone into their fashioning, they deserve better treatment than this. For years I have tried to persuade show societies to stage stick classes in a proper setting. This is one thing for which I admire the Great Yorkshire, where there is lots of space, with numbered sticks set tidily in racks. It is infuriating to lose sticks after a show, but this does happen. I have known them to be thrown in a corner if the owner could not be on the spot to collect them, as though they were bits of timber discarded till next year. Young and novice members, on whom the future of the craft depends, should receive the same consideration as the winners.

When I am judging, the first thing I do is to contact the steward or attendant members. I then expect to be made familiar with the classification and the number of entries in each class. On occasion I have become a bit "worked up" at this stage, for the responsibility of judging 100 to 150 very good sticks, representing thousands of hours' effort, is no light one. I know that I must discard a lot of well-fashioned crooks and sticks, but now have in mind's eye just what I am looking for.

Sometimes there are two judges. Unfortunately the stick which suits one man does not always suit another, so that to decide the order of placing is easier said than done.

I handle every stick when judging. Once I saw a Border farmer who had been asked to officiate. He stood with one hand in a pocket and pointed to the ones of his choice without moving! The balance of a stick is vital and you can no more judge this without handling than a ridden hunter class may be placed without a turn in the saddle. Your hand should slip easily into the neck of the stick. If it will not, neither will a sheep's head, so what is the use of the crook?

12. Leg Crooks and Thumb Sticks

A TYPE seldom carried by any but the working shepherd is the leg crook, or leg cleek as it often known in Scotland. This has a narrow mouth, just wide enough to take a sheep's hind leg, opening slightly so that the leg above the hock is not strained or damaged. Border shepherds swear by these leg crooks, which have the advantage that a sheep may be easily caught when facing inwards and trying to push its way further into the flock. To use a neck crook in such circumstances is not always easy, owing to the pressure of other animals against the one you wish to catch.

Width of head varies according to class of sheep, but an old penny is a general guide for mouth width. Leg crooks may be in wood or horn, and the Royal Highland and Great Yorkshire are but two shows which deservedly assign special classes to them.

Thumb sticks need special care in selection on tree or bush. Their two prongs should be as even as possible, leaving the shank at similar angles. Ash and hazel are among the likeliest sources of thumb sticks.

13. Carving Ornaments

by Frank Graham

IN ADDITION to making crooks and walking sticks, I have been carving ornaments from wood and horn for over 30 years. In essence, the art is the same as making a fancy stick, the difference being that the piece is complete in itself and will not be shanked, nor of course will it be used except for decoration.

Wooden chains are one of my specialities. The whole is cut from from the same block of wood; there is no breaking off or glueing. Beech is useful for this purpose, being softer than elm although without the wonderful finish of the latter.

If your links are to be one inch wide, choose a block of wood slightly more than one inch square in section, to allow for taking some off. The wood should be about a foot long, or as long as you desire your finished chain. Don't be too ambitious to start with, as you will surely have some breakages which mean discarding the whole thing, and the longer the chain, the more difficult to avoid these. At the end of your block, draw a Maltese cross about three-eighths of an inch wide, representing the width of your links. Join the design along all four sides, and take out these corner pieces with saw and chisel. You are left with the block shaped ready for working.

Now measure off the links on all four sides. They will be 1½ to two inches long, although I plan to make a chain like the old shoulder chains on a cart, with big links to start and finish, and smaller ones in between. A penknife is the main tool. Start whittling away when the outline in completed in pencil, and finish each link with a small file.

You will find that as each link is completed, another one is half done. The biggest snags are some unsoundness or unexpected knots in the wood. Beech containing small black marks is very attractive, but especially liable to faults. One local timberman scoured the district seeking beech of this type.

Leave just enough room so that each link turns round. As the chain grows, so does the fascination of making it, but difficulties increase. The new links cannot be discarded or pushed away; they are there, and every movement entails turning over the whole chain as far as you have made it. My chain took 32 hours work; it is the only piece I have timed.

Attractive ornaments may be made from horn. One big advantage is that horns not big enough to make a stick head may be brought into play. Ram's horn with a black line down it shows up well. Goat horn is useful. It is also

scarce, more and more goats being bred hornless or de-horned as kids. It is not quite so good to work as ram's horn. I made some black eagles out of goat horn, etching the outlines with a file as pencil would not show. The front of a goat horn is thicker, termed the "bead" in stick dressing. From this is fashioned the bird's body, with the shelly sides of the horn making the wings. Cow horn may be used, but will not stand heat and will not bend.

Several birds or animals may be carved on the same piece of ram's horn. It is heated, and closed into a smaller circle. The carvings are done one at a time, and a good horn may be slit to give enough "cloth" to make another animal or bird at a different angle.

These carvings may be done using either the point or the base of the horn as the eventual stand. Take a good, solid horn, but not necessarily a long one, and heat the part to be turned very, very slowly, using a funnel lamp. Take great care not to burn the part to be carved. Turn it into a spiral, then bend the upright so that it is balanced exactly in the centre. If you neglect this part, no amount of fancy carving will make a first-class object. A little of the inner edge may be rasped away to make bending easier, otherwise leave the horn in the natural state at this stage. To remove more increases the danger of burning what will be required.

You are now ready to begin your design. Fix firmly by the base in a good vice. Start at the top, rasping down to the rough outline of your model. When you have got fairly near to where you want to be, sketch in your fox, bird or dog, or whatever you choose. Unlike a walking stick, the top outline must be done first. Taking the fox as an example, carve its top outline and then round the head part. Leave the tail in a solid block. Work off in front of the fore legs, then behind them, still leaving the tail solid. The reason for this is that projections like the tail are easily snapped off, and I ruined several carvings at a late stage through trying to carve the tail too soon.

Do all the body work as soon as you are able. It doesn't do to leave a lot of body carving after the legs are finished, because like the tail they have little strength. If I snap off any part, I throw the whole lot on one side. Other people might never know if it was glued on again, but I should, every time I looked at it. This is why I don't like painting ornaments; some may think the paints hides a fault.

All this time the lower part has been fixed firmly in a vice, so it was left rough. Finish by rasp and file working down towards the vice, and altering the position of the grip until you come to the end. If after finishing you decide that the angle is not quite right after all, heat again very mildly, possibly using a candle. A lot of hours are spent finishing off, using file and

Carvings in wood and horn. Top: the eagle (left) is from black goat's horn, the horse on the right from a wooden block, and the foxes next to it from a single ram's horn. Centre: A set of carvings which includes a black and white sheepdog. Bottom: Chain fashioned from a block of wood, showing the links and the way the block should be shaped.

Yorkshire farmer, Mr. Frank Graham, shows carvings in wood and horn.

polish as described in finishing off sticks. This final polish makes all the difference between a second-class article and something of which you can be really proud.

Animals and birds may be carved from wooden blocks. Otter, fox, mare and foal, dog, pony are among my favourite subjects. English yew is very suitable timber, but like beech it doesn't have quite that marvellous grain that elm possesses when polished. Yew is softer than elm, however.

Your block of wood must be dry. Allow for a good firm base, and sketch in the outline above the base. Don't forget that horse's belly is wider than its shoulders, and so if you do not leave plenty of wood at this point you will finish with a flat-looking animal. I have done this, and had to learn through my own mistakes; far better start with enough wood in the right places. On another occasion I carved a fox with his head looking sideways. Then I found that I hadn't enough wood for his nose, so I had to swing it back again.

Having sketched your outline, start work with brace and bit. Bore out as much as you can, then set to work with small file or penknife. Work all the time with the grain of course, or you will soon be in trouble. Final polish is helped by polyurethane, as with horn. Best of all for the very last stages is just a cloth. It takes a tremendous lot of time, but gives a great feeling of satisfaction. I will not attempt to describe the satisfaction of wood and horn carving. It is something that grows, so that the further you advance the more fascinated you become with your work. All you need are a few blocks of wood or horns, and very simple tools. At the end you have something really worthwhile. There is not much gained watching television night after night.

Glossary

HEAD — The part grasped by the hand. It may be of horn, wood from a separate block, or a continuation of the shank.

SHANK — The stem or handle, below the head, made from any one of a variety of sticks, e.g. hazel, ash, holly.

FERRULE — Either the strengthening joint between shank and head, or the protecting cover of the tip.

MOUTH — Opening of the head, which in the case of a crook takes the sheep's neck or leg.

HEEL — That part of the head in line with the shank. It is shaped into either a square or a round heel.

NOSE — The culminating point of the head. It may be fashioned into a decoration, but it should never drop below the level of the base of the heel.

CROWN — The highest point on the head. It may be central or slightly to one side.

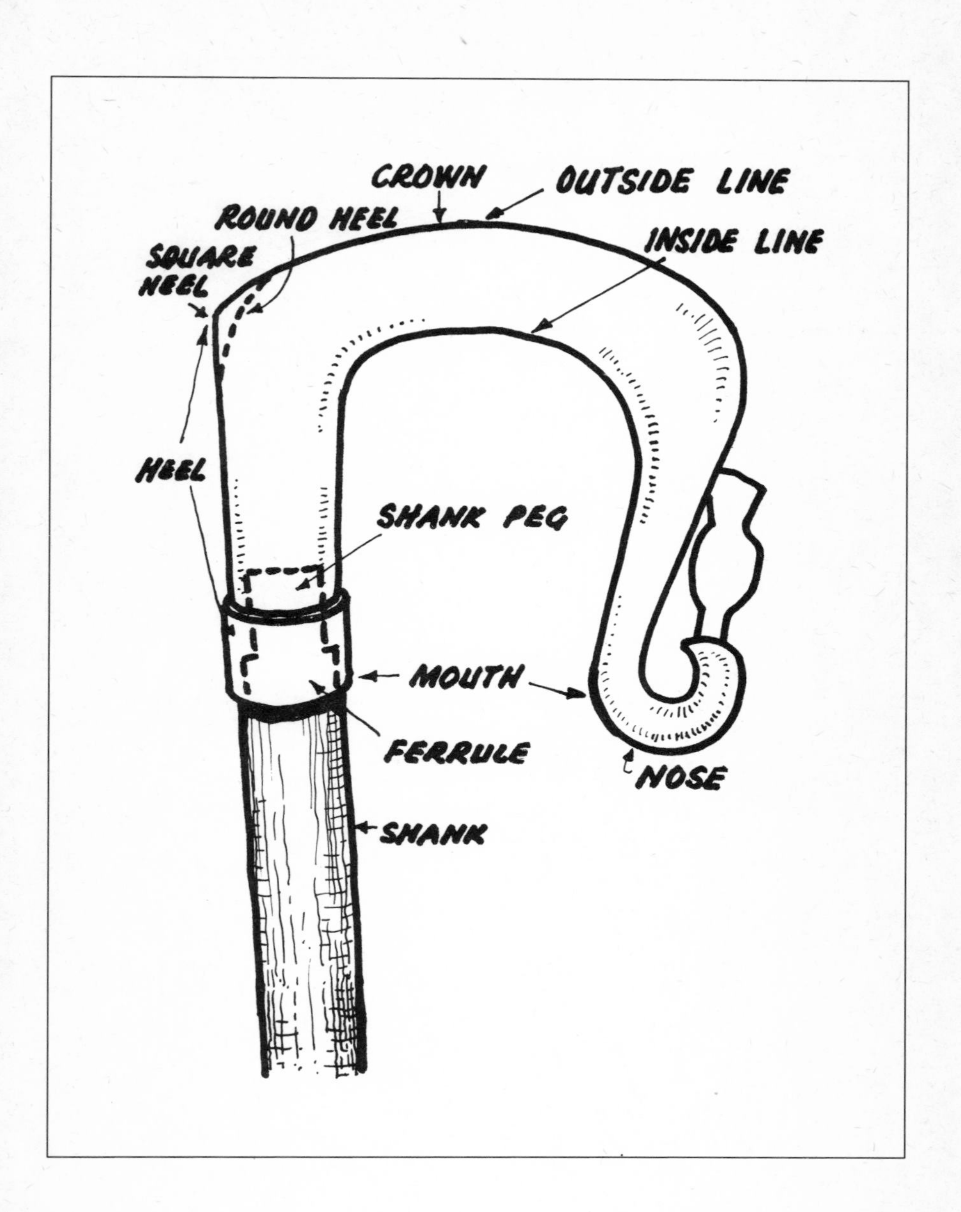
CROWN
OUTSIDE LINE
ROUND HEEL
INSIDE LINE
SQUARE HEEL
HEEL
SHANK PEG
MOUTH
FERRULE
NOSE
SHANK